Lea Utsira's

All About
Leaves

A Science Mindset Book

Photographed and
written by Lea Utsira

Lea Utsira's

All About Leaves

A Science Mindset Book

What do leaves do?

Leaves have many jobs. These leaves are on a plant called Lady's Mantle. They are green.

Do you see how the raindrops stay on top?

The leaves on this plant are also green, with white edges. This plant is a vine called Vinca.

Do you see how it stayed green, even in winter?

Leaves are remarkable. There are so many things to observe about leaves.

The leaves here are not green. These Maple leaves are yellow.

In Summer, they were green. In Autumn, when the days grew shorter, and the air was colder, they changed color.

These long, narrow leaves have another job. They poke holes and grow up through the papery, brown carpet of dried, fallen leaves.

They are the earliest Spring flowers, and sometimes come up when snow is still on the ground.

These plants are Snowdrops.

The flower buds have set, soon to open into little, white flowers. New leaves are poking through, again!

Little flowers have started blooming!

The leaves absorb the sunlight and grow the Snowdrops bigger. The roots of Snowdrops grow bigger too. The energy of the sun travels down to the little bulbs to make more.

Snowdrops are spreading all over! The narrow leaves capture the energy from the sun. The Snowdrops grow and spread.

This is a Maple leaf in Winter.

It froze in the water.

When leaves freeze, they break down.

The water expands when it freezes, and that breaks down the cells of the leaves.

Later it will completely break down and become soil.

Light shines on and through leaves. This plant is a Wild Geranium, called Herb Robert. Light from the sun is absorbed by leaves. Light is energy. The leaves take the energy into the leaf colors in pigment cells, or color cells. What happens is photosynthesis.

Trees grow taller from the sun's energy that the leaves
have captured and transformed. The stems and
branches and tree trunks are all there, growing,
because of the leaves. In photosynthesis, light shines on
the leaves' pigment cells, and the light is converted to
energy for the plant or tree to grow.

This is moss. Moss is green, and it grows. It also takes sunlight to grow. But moss does not have leaves. Moss is like a carpet of green stems that look like leaves, but are not.

Lavender has gray-green leaves. They have a strong smell. Some leaves, like Mint, Basil and Parsley, have a distinctive fragrance. You can smell it on your fingers after rubbing and crushing the leaves in your hands.

The leaves of Ferns have another job. Ferns do not grow flowers, or make seeds. New Ferns grow from spores on the leaves. Spores are tiny bumps that develop on the underside of the Fern leaves. They will grow into the new Fern plants.

Many trees drop their leaves in fall. They go dormant for the cold winter. This means they do not take the energy from their leaves, or grow, during the cold season. Trees that drop their leaves and go dormant are called deciduous trees.

There is another kind of tree that stays green in winter. This kind of tree is coniferous, or a conifer. We also call them evergreens. They have narrow needles for leaves. These needles collect the sunlight for energy to grow, the same as for deciduous, or broadleaf trees.

The big tree here is an Oak. The leaves are starting to change from green to yellow, in early Autumn. As the leaves fall, more light shines through the woods. When the leaves are green and thick in the middle of Summer, the woods are dark and mysterious. The leaf cover at the tops of the trees is called the Canopy, like a tent.

Oak trees provide acorns as food to animals in the forest. We must take care of our Oaks, our mighty Oaks!

This is an Oak leaf in the fall. This big, brown leaf has veins going from the tips into the center. Through these veins flows the energy the leaf has captured from the sun. Oak leaves break down, or decompose, into very rich soil, after falling from the tree. We call leaves that have become soil, leaf mulch, or compost. This return to Earth to become soil completes the cycle of Spring growth, Summer sun capture, Autumn color change, and Winter dormancy.

Seeds from the trees can fall and start growing right where the leaves fell before. Leaves fall and become the soil for the next generation of trees. Growing new trees is also one of the jobs of leaves.

Leaves catch sunlight! Now you know all about leaves.

Glossary

canopy~the top layer of trees form the canopy of the forest

cell~small, closed compartment of leaf structure, similar to a bubble

coniferous~a type of tree with needles that stay green all year long

deciduous~a broadleaf type of tree, which lose the leaves in the cold season

decompose~the breaking apart of natural material by wear and tear outside

dormant~a phase in the life of a plant or tree when growth does not occur

pigment~coloring matter in the cells of leaves

photosynthesis~the transformation of energy from the sun to sugars in plants

spore~small parts of mosses, ferns and algae which act like seeds to reproduce

Lea Utsira taught Early Grade Science, Kindergarten and Pre-K for more than 20 years. She now maintains a little woods in the Hudson Valley. She gardens, photographs and writes for children every day. Lea thanks all of the young children who observe nature, show curiosity, ask questions, and inspire everyone in their world.

Thank you! Takk for det!

What do leaves do? Observe leaves in all seasons, and find out with Lea! Look, think and learn.

All About Leaves

iUniverse books may be ordered through booksellers or by contacting:

iUniverse
1663 Liberty Drive
Bloomington, IN 47403
www.iuniverse.com
844-349-9409

ISBN: 978-1-6632-2040-0 (sc)
ISBN: 978-1-6632-2067-7 (e)

Library of Congress Control Number: 2021906636

Print information available on the last page.

iUniverse rev. date: 04/01/2021